ELECTRIC CARS WILL DIE IN 2027

ELECTRIC CARS WILL DIE IN 2027

Masafumi OYAMA

ISBN: 978-1-64273-621-2

Author: Masafumi OYAMA
Projct Editor: Yumi Itabashi, Hachidori Inc.
Editor: Andrew Kirkham

Printed in Japan
1 2 3 4 5 6 7 8 9 10

43-32 22nd Street STE 204 Long Island City New York 11101
www.onepeacebooks.com

Introduction

In 2025, when Republican Donald Trump returned to the White House, the world was in a big uproar about two other top news stories in the global automotive industry. The first was the rumored debut of Toyota Motor Corporation's FT-Bh, an ultra-fuel-efficient hybrid car boasting a range of 50 km per liter, first unveiled as a concept car at the 2012 Geneva International Motor Show. The rumor generated quite a buzz. The other story was that BYD, a Chinese electric vehicle manufacturer, has developed an innovative platform called the "Super e-Platform," enabling high-speed charging for EVs (electric vehicles). This platform allows achieving a 400 km range in just 5 minutes. So, what do you think and how do you respond to these infesting international headlines?

My genuine feeling about the first 50 km/L hybrid car was one of astonishment. I was like, "It's absolutely an amazing breakthrough in terms of automotive fuel efficiency!" I was fascinated by the relentless effort of those who contributed to the automotive industry, which led to significant improvements in fuel efficiency. I was also convinced that the very goal of net-zero CO_2 emissions by 2050 was achievable if the trend continues in that direction. Automobiles can definitely achieve one liter per 100km. As you are all well aware, our current world economy is focused on the issue of global warming. Nearly all major corporations and research institutions worldwide are in fierce competition in research and development over one question: "How do we reduce greenhouse gases?" Undoubtedly, improving automobile fuel efficiency is a significant step in fighting the climate crisis. I should be, and I am more than just happy to see the big news.

My impression was, yet, 180 degrees different when I saw another breaking world headline BYD made about the significant reduction in charging time. The long charging time had been the biggest issue with EVs (electric vehicles) when compared with those conventional natural gas vehicles with an internal combustion engine. I understand that a regular household outlet can never deliver such a high level of power, so I didn't brace myself for such a rapid and significant advancement in EV technology. I used to believe that emerging EV technologies would be around someday, I don't know, probably in another decade or two, but the headline said no.

I didn't expect to see it come this fast or progress this quickly, so the news sounded like a tech warning that it is going to be a rough ride.

"Could it be that gas-free vehicles dominate the market and replace all traditional options? If so, will China's automobile manufacturers kill off all other competitors soon?" Let me tell you, honestly, the impacts will not be limited to the big three automakers in the US. Toyota is outranking the competition in this unprecedented era of fuel efficiency, but it's not the only one to be affected. Even Tesla, which has long pioneered the industry in EV sales, will be drawing near to its time under the momentum of China's automotive manufacturers.

In other words, the bare fact is that the dedication the Chinese government has consistently pursued to crush the industries of other advanced countries has finally paid off.

As you know, times have changed, and China's economic evolution has become a stiff headwind for us in the West. At the same time, the country has never hidden its ambitions for global hegemony. Particularly problematic among the string of issues are the continued expansion of her beautiful bill of defense budget (which reached double digits) and the economic hegemony initiative commonly known as the "One Belt, One Road".

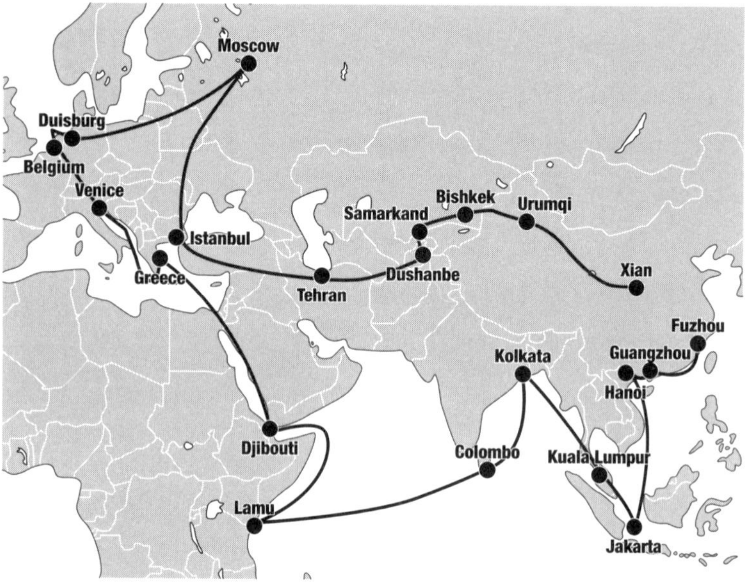

Figure 1 The Map of "One Belt, One Road"

"One Belt, One Road," or "The Belt and Road Initiative," refers to China's primary national strategy for economic reform, announced by Xi Jinping upon his inauguration as president in 2013. It especially envisions a vast regional economic zone modeled after the ancient Silk Road connecting China with Europe, linked by land routes through Central Asia and sea routes via the Indian Ocean. The model was achieved through the development of transportation infrastructure, such as railways and ports. Since China has already become the world's factory, the initiatives aim to expand the global export network for Chinese products, thereby displacing other countries' industries and enabling China to weaponize its economic power in support of its unnamed global hegemony.

Under this grand vision lies the "Made in China 2025" strategy, designed for the efficient development of domestic industries with the core objective to take Chinese sectors to the new heights meeting the highest standards over the globe through excessive state-directed protection and massive subsidies, focusing primarily on technologies to cope with CO_2 emissions—the central counter-measure against global warming, currently the most severe cause of environmental destruction. (The US government has defined this as "China's state capitalist system")

The momentum of this dirigisme is now unstoppable, as unveiled by the headlines about BYD's "significant reduction in the charging time for electric vehicles". This is because, within both "Made in China 2025" and the "One Belt, One Road" initiative itself, the sector at the very heart of development and industrial cultivation is none other than the electric vehicle industry.

The technology for shortening EV charging times made me acknowledge, once again, how petrifying the threats posed by the country are. It is not merely a date for a not-so-soon-to-be-expected event. The global strategy crafted by President Xi is well thought out. The "One Belt, One Road" initiative demonstrates China's deep understanding of the current worldwide situation and prospects, China's geographical characteristics, and the technical challenges faced by Chinese industries. The policy outlines how to address all these issues to create a strong culture of repeatable success, thereby managing geopolitical threats, winning over competitors, and ultimately enabling China to seize hegemony over the global economy. This strategy is well-curated.

Of course, not everything in the initiatives went as planned. China's economy is on the cusp of a deflationary death spiral. The economy is apparently growing, but cracks are showing.

An accurate depiction is, for instance, the historic population decline (a deepening birthrate and fertility rate crisis) in recent years. Before COVID, China's birth rate had been steady at around 1.50 children per woman, but it then became a demographic disaster after the pandemic, falling sharply to just 1.16 in less than two years.

As you know, the "special total fertility rate (TFR)" is the average number of children a woman has over her lifetime, calculated by dividing the total number of children by the total number of women of childbearing age. Hence, it is rather doubtful that the number of births would drop this significantly just because the birth rate declined slightly for a year or two. That means these figures tell the story of how China's birthrate has been crashing since COVID-19.

Given that I have researched Japan's declining birthrate and aging population for years, I am convinced that I have deep insights into this subject. The key determining factor in Japan's birthrate is, in fact, "the tax Japanese pay on average yearly". When this figure has risen in the name of fiscal consolidation, Japan's birthrate has accordingly plunged. Notably, Japan has endured much during the lost decades, including economic failure; in all, the average laborer's pay has barely increased. Consequently, as the tax burden on citizens grows heavier, their take-home pay decreases proportionally. Those various deductions reduce Japanese people's real wages, leading them to have fewer children.

Economic issues alone, of course, do not necessarily explain the decline in birth rates. Yet, given how China's birth rate has plummeted so dramatically, and the population is shrinking in less than two years, the only conclusion is that Chinese living standards have rapidly declined. This hidden collapse is the actual, real reason.

What would happen if there were a decline in livability in another country? Yes, riots would likely have erupted long ago. Chinese people did not riot because their government is so intimidating, or rather, Xi Jinping, the dictator, is intimidating… That is the way fear manifests in China.

Unlike my American fellows, I have been living my life very much closer to China, and thus I have been constantly and directly feeling how terrifying the country is. Looking back on history, it is deviously haggling, highly calculative, cruel, and above all, utterly inhuman. Its government has long adopted a "social credit system" (a system that provides a holistic assessment of an individual's or a company's trustworthiness through the internet, surveillance cameras, and other means, as an extension of existing social rankings and ratings) to monitor its citizens' lives. It has persecuted a huge number of those who have voiced dissent against the Chinese Communist Party. Since the government is so committed to silencing the voices of opposition, the social credit score has become a source of terror for all its citizens. The Chinese government is leading a rite of passage of such "state capitalist system", even at the cost of creating a brutal crackdown on its own people. US manufacturing was then killed before anyone realized.

Having observed the global economy for years, I've come to realize that doing business with China is only profitable in the short term. All countries and all corporations suffer devastating losses at the end, without a single exception.

The American auto industry, including Tesla, which recently built a large factory in China, is pledging to invest billions in electric vehicles that aren't selling. The current internet connections between Western countries and China have been blocked, resulting in excess losses for IT agencies, including the five Big Techs (Apple, Alphabet, Microsoft, Meta, and Amazon). These firms are all victims that have the most money in China to date. When we take a closer look at the facts, we see that the corporations that grappled and invested the most in China suffered the greatest losses. This is not limited to US firms. Looking around the world, from Japan's SoftBank and Rakuten to Nidec, and even Samsung in South Korea, the adverse effects are spreading far and wide across the globe. And before we knew it, the dominance of the dollar triggered by the Russia-Ukraine war had already flooded the American market with cheap products from China.

Despite appearing superficially stable, the U.S. economy is unfortunately in a state in which American industries are being forced into an unfavorable position due to the complete failure of the previous Biden administration's policy on the containment of the adversary.

We were all under the delusion that we could rake in the cash by doing business with China… but it was our merest, excessive daydreaming. This entire electric vehicle story was just about how we constructed our own reality in which China would make it rain if we worked hard to produce electric vehicles. So, what is the reality? As illustrated by the recent slump in sales to China by the Big Three, the dream crumbled like a house of cards. We would eventually be forced to abandon the internal combustion gasoline engine technology we had relied on for so long and switch to EVs (electric vehicles), flush money away to make the auto businesses in China profitable.

In 2024, former President Donald Trump was elected to a second term in the Oval Office, marking a turning point in this China-oriented electric vehicle landscape. Trump reassessed the various connections with China and Russia to address Biden's blunder and failed presidency. One of the new initiatives, immediately after he took back the White House, was the imposition of substantial tariffs and a hefty tax hike on countries trading with the United States. The WTO (World Trade Organization) should govern the rules of the global capitalist economy and create an international legal framework to ensure smooth business exchange. Still, it had already become a dysfunctional, paralyzed colossus because of China's cover-up.

Now, what? More and more countries hit each other with reciprocal tariffs to initiate discussions with China outside the WTO. Personally, I foresee the situation unfolding like this-

"Everyone believes that it's time to execute the ultimate countermeasure against China's economic hegemony... electric vehicles have long been regarded as the solution in the next phase of transportation to replace natural gas vehicles that spew carbon dioxide. But no. Let's stop, clear our minds for a second, and think again..."

It's impossible to overturn it now, even with a stipulation from the US president, because electric vehicles are well on their way to mass adoption. Still, I genuinely believe that electric cars will die in 2027. It is not merely a prophecy. No. Current figures and global issues serve as the basis and evidence. Our world is waiting for ignition, and then everything will change.

The trigger will come in 2027, several months before the National People's Congress, which is held every five years, in China, when Chinese naval vessels will suddenly enter the waters near the Senkaku Islands, dozens of kilometers from Taiwan, encircling and blockading the island. And the moment the news will spread and shock the people all over the world, the fleet's guns will roar one after another as a bugle call-the beginning of China's military invasion of Taiwan.

Then, electric cars will die…

Table of Contents

Introduction 1

Table of Contents 13

Chapter 1 The far east region under China's military threat ... 15

1. When China's military invades Taiwan 16
2. China for South Korea 18
3. President Obama's China containment policy 19
4. The "comfort women agreement tossed about by China and the US 23
5. Why does China want Taiwan this badly? 28
6. Economic concerns 33
7. China's economic gloom mounts 36
8. World history will change in just 15 years 48

Chapter 2 China's coercive economic tactics 57

1. The background of electric vehicle promotion 58
2. China's automotive strategy —taking advantage of the Paris agreement 64
3. Global EV outlook 66
4. Joe Biden's failed presidency 71
5. EV reliability Issues 75

Chapter 3 The way EV dies 83

1. The biggest blunders of global warming policies 84
2. The rapidly advancing carbon capture technology 85
3. New ways to utilize recycled carbon dioxide 87
4. Chemical fuels that make EV unnecessary 88
5. President Trump's China strategy 93
6. The national people's congress in 2028 87

Chapter 4 The terrifying ending 100

Afterthoughts 105

Author profile 109

Chapter 1

The far east region under China's military threat

1. When China's military invades Taiwan

In April 2025, the US Senate confirmed the nomination of Elbridge Colby as Under Secretary of Defense for Policy, making him the principal adviser to the Secretary of Defense on shaping national security policy. He has long criticized Asian countries, especially Japan, for having insufficient defense spending. He has also urged Japan to raise its defense budget from the current minimum of 1% of GDP to at least 3%. This is not his merest, subjective opinion. Former German Chancellor Merkel, who was visiting Japan at the time, also confirmed Japan's defense budget was in the crosshairs.

Setting aside the harsh realities of Japan's economy, my genuine response to the fact that eyes turned to Tokyo was rather like a keen awareness of the high risk of trouble in the Taiwan Strait. Personally, I came to the full realization that Japan must raise its defense spending as a preventative measure as well.

The deciding factor prompting senior government officials in Western nations to urge Japan to increase military spending is indisputably South Korea's 2025 presidential election, in which Lee Jae-myung of the Democratic Party of Korea was elected president.

The inauguration of the new president set up the stage for China's preparations for the invasion of Taiwan.

Now, here's a question for people in the USA. What do the Japanese really think of the United States and the American people? Do most Japanese like and respect the USA as the world leader? Is Japan simply taking advantage of the US based on the US-Japan Security Treaty? Or are Americans nothing more than just business partners to the Japanese? Perhaps the Japanese still hold a never-changing grudge against America as the enemy that dropped the atomic bomb?

We, Japanese people, do feel thankful for Americans (be mindful, I am not speaking on behalf of the new generation). Thanks to the tremendous support from the US government, the Japanese economy recovered remarkably from tatters after WWII. The support mainly consisted of purchasing our products, which were cheap but of poor quality at the time, and that is why we, the older generations in Japan, love Americans.

Here comes the question-who do the Koreans appreciate then? The fact is, there is a large portion of the Korean population that harbors such gratitude not only toward Americans who stuck together with them through thick and thin in the Korean War, but also toward the Chinese. And the political party that upholds a more pragmatic, amicable approach to China and plans to foster friendly relations with it is the Democratic Party of Korea, led by Lee Jae-myung, the new left-leaning president.

2. China for South Korea

The economic disaster that almost brought South Korea to its knees was the 1997 Asian financial crisis... The memories of the period are still fresh in my mind. I can't forget the shocking headlines like "for economic recovery, the South Korean government would have to sacrifice its people for the foreseeable future." The analysis was largely accurate because Korea lacked sufficient foreign currency reserves to repay its foreign-currency-denominated debts and rebuild the economy. Imports would decrease if the Korean economy were to decline, leading to a trade surplus. Achieving a trade surplus would, in turn, generate foreign currency reserves commensurate with that surplus.

However, tracing the subsequent trajectory of the South Korean economy reveals that the widely predicted economic deterioration did not materialize. Contrary to expectations, the economy steadily recovered, and so did the foreign exchange reserves. By now, you might probably have a lovely idea of what happened-it was China that chipped in to rescue South Korea at that time. By expanding exports to China, South Korea grew and rebounded from contractions. In other words, China did something to Korea like what the USA did to Japan after WWII.

As mentioned, the problem here is that the more deeply a country or a company is economically tied to China, the greater the damage it will suffer afterward. For South Korea, that turning point came during the Park Geun-hye administration in 2016.

3. President Obama's China containment policy

In 2013, when Xi Jinping rose to power and became China's president, a storm was on the horizon over China's previously steady progress toward democratization and market liberalization. Feuds were escalated between China and its neighboring countries. For example, while China had long asserted sweeping claims to sovereignty over land parcels and their adjacent waters of the South China Sea, it was only after Xu's inauguration that China began committing such outrageous acts as arbitrarily island-building, naval patrols, and construction of military bases in the disputed region, disregarding the United Nations Convention on the Law of the Sea completely.

On top of the dispute over territory in the South China Sea, there is the international issue of "one country, two systems," which describes the governance of Hong Kong. China made a declaration, signed with the UK, stating that "Hong Kong is granted autonomy for 50 years after the handover without any changes," an empty promise. Not only did China forcibly impose the National Security Law on Hong Kong, but it is also hardening its stance against dissent, as in the combative "Wolf warrior diplomacy," including the repression of Uyghurs in Xinjiang. It backfired, resulting in ever-growing resentment.

In response to such issues, President Obama, the US president at the time, decided to implement four countermeasures as part of his policy toward China. First, through the TPP (Trans-Pacific Partnership Agreement), the United States and other nations centered around the

Pacific created economic integration to negotiate collectively with China and counter its economic pressure exerted on individual countries.

Second, following the collapse of the Soviet Union, the United States anticipated a contingency involving Taiwan and reestablished its military alliance with the Philippines, a country from which US forces had temporarily withdrawn, to ensure that US forces could continue to use bases there supporting the US armed forces.

Third, for the same reasons above, Obama decided to deploy the Terminal High Altitude Area Defense (THAAD) system in South Korea in 2017 to strengthen deterrence against China in the Far East Asia region.

Finally, the conclusion of the GSOMIA (General Security of Military Information Agreement) between South Korea and Japan, which had not yet become military allies at that time, ensured that the US Forces in Japan and Korea, along with the governments of Japan, the US, and South Korea, could swiftly share information in the event of a crisis in Taiwan. This facilitated streamlined military operations among the nations should the need arise. This was implemented exactly during Park Geun-hye's administration in South Korea.

It was evident that the series of policies provoked serious opposition from the Chinese government. In particular, the radar deployed in South Korea as part of THAAD has a very long detection range, allowing the US government to visualize the movements of military facilities in eastern and northeastern China. The Chinese government thus began exerting pressure on the South Korean government through every possible means to prevent the deployment of THAAD at all costs. One example of this was the boycott of South Korean products in China around that time.

As captioned, China rescued South Korea, which was teetering on the edge of national bankruptcy due to the "Asian financial crisis" in 1997. Given these circumstances, there is no doubt that the boycott exerted visible and invisible pressure on the South Korean government. Consequently, ever since then, whenever China and the United States clash in the second Cold War, South Korea has remained indecisive and struggled to take a side.

Economic pressure can hardly stand alone to completely ruin the relationship between the United States and South Korea. The Chinese government, therefore, devised a plan to replace South Korea's president and parliament with a political party under its influence. The

Democratic Party of Korea was then selected. Back then, it was led by Moon Jae-in, who had a relatively close relationship with North Korea. In 2017, when Moon Jae-in served as the 12th president of South Korea with China's backing, he promptly began scrapping all the military agreements reached with the United States during the Park Geun-hye administration to improve relations with China. He introduced "Three Noes" and stated that he would halt additional THAAD deployments, withdraw from the US missile defense system, as well as the military alliance with South Korea, the US, and Japan; he also delayed THAAD deployment time after time despite the repeated strong messages from the US. In fact, he has never nodded to any of the demands from the White House. In 2019, he even terminated the extension of the GSOMIA (General Security of Military Information Agreement), a bilateral military treaty concluded with Japan for the exchange of military information.

4. The "comfort women agreement tossed about by China and the US

Here, we should note in particular that China has skillfully embedded overlooked historical issues between Japan and South Korea to undermine both the THAAD (Terminal High Altitude Area Defense) system and the GSOMIA (General Security of Military Information Agreement).

Initially, it was former President Obama who worked to mediate between Japan and South Korea (the two nations remain at intense odds) to secure the signing of GSOMIA (the General Security of Military Information Agreement) between them. In Japan, where I live, there was a prevailing sentiment that the public was fed up with the South Korean government constantly making promises to Japan that turned out to be lies.

The same holds for many South Koreans. Many of them cannot get over the humiliation they endured under Japanese rule in the colonial era. However, to enable Japan, the United States, and South Korea to respond swiftly in the event of a Taiwan contingency, it is vitally important to improve Japan-South Korea relations and conclude the GSOMIA between the two countries.

However, the Park Geun-hye administration at that time continued to wallow in the negativity of Japan's military invasion that occurred as long as 70 years ago, and forced it upon other countries when she toured around. Those actions proved that Korea is trapped in a state of victimization that sparked the row with Japan. President Obama then stepped in and asked for reconciliation between the two countries through resolving the issue of wartime comfort women, which remains a lingering problem in postwar reparations. The offset laid the groundwork for the conclusion of GSOMIA. (The "comfort women issue" concerns whether postwar compensation for comfort women should be covered under the modern Japan-South Korea Claims Settlement Agreement or excluded from it, as international law stipulates that inhumane crimes committed during wartime fall outside the scope of agreements determined by intergovernmental consensus.)

The "Comfort Women Agreement" was then concluded between then-Japanese Prime Minister Shinzo Abe and South Korean President Park Geun-hye in 2015, stipulating that in exchange for the Japan's expression of "apology and remorse" and the contribution of compensation to the victims, South Korean shall also "confirmed that this issue has been resolved finally and irreversibly" (so the South Korean government shall never revisit the issue). This eventually paved the way for the successful conclusion of GSOMIA between the two countries in 2016.

Nevertheless, this trilateral relationship constitutes a significant obstacle to China's invasion of Taiwan, without a shadow of doubt. China, therefore, decided to work behind the scenes to side with South Korea and stir up dissatisfaction with the "comfort women agreement" again, aiming to create a crack in the relationship among South Korea, the US, and Japan. And it was Moon Jae-in, the favored candidate of China, who stepped up as South Korea's president in 2017.

Moon started his presidency timeline overturning the Japan-South Korea "comfort women agreement" as he claimed it was an emotional invalidation for the victims. He moved on to provoke the 2018 incident where a South Korean Navy destroyer locked its fire-control radar onto a Japanese Self-Defense Forces patrol aircraft over the Sea of Japan. To top it all off, regarding the long-standing issue of forced laborers between Japan and South Korea, he appointed a Supreme Court Chief Justice with ties to the Democratic Party of Korea, who then issued a ruling recognizing compensation claims against Japanese corporations from the wartime period. In fact, the issues between the two countries were all initiated by China, which is why THAAD and GSOMIA were off.

The real issue here is not "is Japan more justified in their claims regarding postwar reparations or is it Korea?" Instead, it is what China did-it cunningly exploited the lingering resentment between these two nations to advance its own global hegemonic ambitions. As you might know, the story then developed as Korea-Japan relations took a sharp downturn after the South Korean government's sudden breakoff. The US, which tried to intervene, was also left completely humiliated. Japan then reinstated South Korea on the Catch-All Controls, and the latter retaliated by terminating the GSOMIA. This is, in fact, exactly what China longed to see.

The storyboard proved that the Democratic Party of Korea was entirely under China's influence. In South Korea, as of 2025, the country is welcoming a new president following the resignation of former pro-American President Yoon Suk-yeol amid a failed internal power struggle. That new president is Lee Jae-myung, the new leader of the "Democratic Party of Korea," previously led by former President Moon Jae-in, the Chinese satellite.

Moon kept his Beijing benefactors close enough for us to expect a gradual reduction in the combat strength of US forces in South Korea in the near future. Put another way, China has already found success in cutting the US troops as the defense force for its preparation for an invasion of Taiwan. Everything is all set, ready for the invasion.

5. Why does China want Taiwan this badly?

The Chinese government aims to seize control of the Far East Asia region, which includes Taiwan, Japan, South Korea, and the Philippines, as the first step toward achieving global hegemony. The country is on a path to dominate the world, seeking to kick out the US as the international hegemon.

No worries, I am not stuck in unrealistic thinking. Why? Tell me, what was the point of smuggling large quantities of the illegal drug "fentanyl" into the USA and making it fragile from internal affairs? Let me say it clearly and honestly, the core of all these issues is that China doesn't see the US even as a threat. China is not afraid to seek global dominance even at the cost of confronting the US. The point is that all these moves aim to cut off relations with Japan, Canada, and Mexico, the countries that are geographically and economically linked to the United States closely, to get a better position in the fight, if at all possible.

I have said it a thousand times, but I will say one more time here- when we look at Xi Jinping, we see a human but no humanity. The "fentanyl" incident is merely a preparation for the inevitable confrontation with the US. China is on the way. Let's bear this in mind.

China sees securing control over the Far East Asia region as its stepping stone to projecting power for global hegemony, and the invasion of Taiwan is its first step, but why? Why is Taiwan so crucial for China?

First, military reasons. The map below shows China's current territorial claims on the world map. As we can see, China is at the heart of the vast Eurasian continent. Despite its enormous size, coastal areas are surprisingly limited. Taiwan, on the contrary, lies at the heart of an arc of islands stretching from Japan and the Philippines down to Borneo. If Beijing were to gain control of the island, it would drastically expand its ability to project naval and air power into the Western Pacific. Surrounding the coastal region of China are the Philippines, Taiwan, Japan, and South Korea. The fact is that among these four countries, only Taiwan does not have a military alliance with the United States.

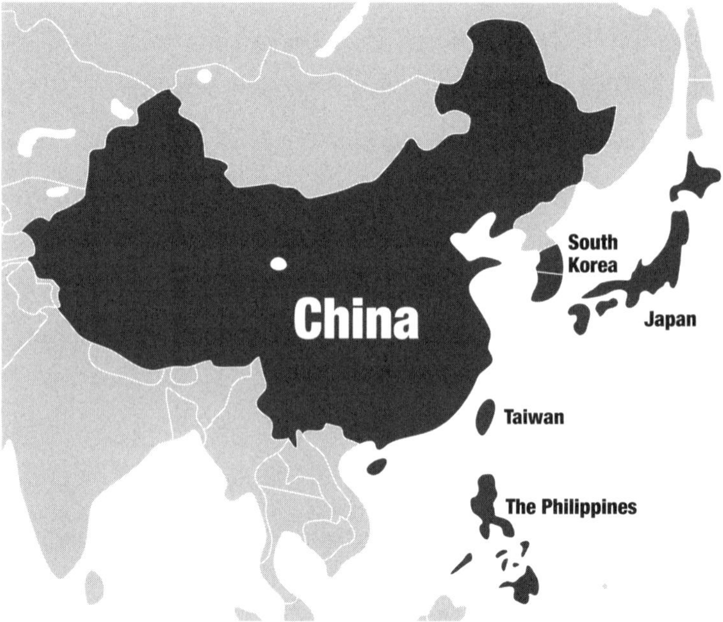

Figure 2 The World Map with China at the Center

The United States military maintains an extensive global infrastructure, operating in South Korea and Japan. It has also signed a military agreement with the Philippines that authorizes US forces access to agreed locations in the Philippines for security cooperation exercises and military training activities. This arrangement serves as the forefront of preserving the free and open regional order, an obstacle to China's military expansion into the Pacific.

Picture if a Chinese submarine were to attempt to enter the Pacific Ocean, it would typically do so by passing through the maritime route between Japan and Taiwan. The submarine must pass through the Senkaku Islands of Japan and be detected by US military radar at the same time. Suppose China intends to flex its power in the Pacific. In that case, it must expand its military influence and gain control over the four countries blocking the maritime route to the Pacific-Taiwan, Japan, the Philippines, and South Korea.

Furthermore, controlling these four countries has significant military-strategic implications for the United States in another sense. This is because severing the military alliances and cooperative relationships among these four countries and the United States would also diminish the influence of its forces in the Middle East.
Suppose that the United States Seventh Fleet, which protects Japan and its surrounding regions, has deployed massive military bases in Pyeongtaek, South Korea, as well as in Yokosuka and Okinawa, Japan, serving as its crucial command centers and supply bases. Rendering these military bases inoperable would significantly degrade the combat capabilities of the Fifth Fleet, which has its area of responsibility in the Middle East, encompassing the waters from the Persian Gulf to the East African coast, and currently operates without dedicated

vessels. Rather, it assembles its forces by deploying ships from the US Navy's Atlantic Fleet and Pacific Fleet as needed. Reducing the combat strength of the Seventh Fleet in the Pacific would, hence, directly lead to a reduction in the Fifth Fleet's force as well.

If the Fifth Fleet's combat readiness plunges, it would significantly undermine China's economic hegemony because the "One Belt, One Road" initiative, the very heart of China's global hegemony strategy, relies on a key maritime route through the Arabian Sea, the Persian Gulf, and the Red Sea to reach the Mediterranean. Expanding China's military influence in this region would enable it, as the world's

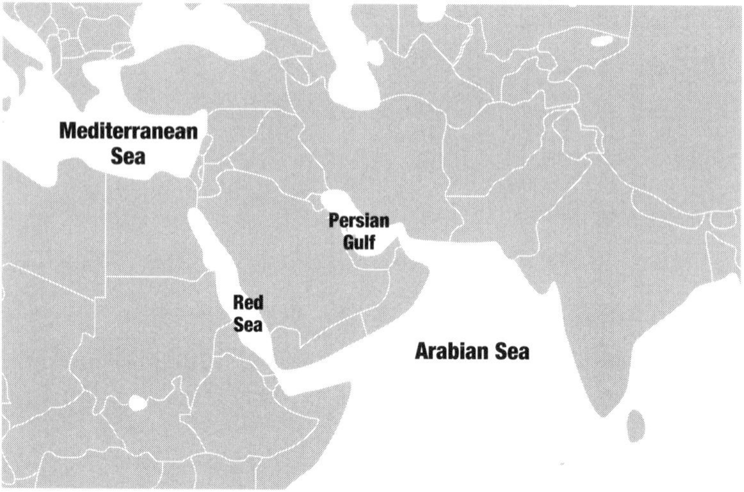

Figure 3 The Nautical Chart of the Middle East region

largest oil importer, to take the maritime route to import large volumes of oil. This, in turn, would allow China to export even more of its manufactured goods worldwide.

To explain the impact of "One Belt, One Road" in the Middle East, China seeks to create a rift between the four countries and the US military, which has formed a barrier along China's coastline.

6. Economic concerns

Another reason is that most of the advances in core semiconductor manufacturing technologies that China has long coveted are owned by Taiwan, Japan, and South Korea.

In recent years, China has repeatedly provoked conflicts with various nations, leveraging its economic and military power. The US, as well as other Western countries, is growing alarmed.

At the beginning of the Biden era, the White House exerted a series of strategic efforts to counter the growing Chinese influence. Two core policies include tightening the restrictions on the export of high-performance semiconductors and related technologies to China, and excluding five corporations from China, including Huawei and ZTE, from internet businesses in the USA-oriented Western world.

The US alliance takes such countermeasures stemming from concerns that China's military power may surpass that of the United States. This is because China's military strength, driven not only by double-digit annual increases in defense spending but also by the potential military application of its most powerful supercomputer and AI technologies in the world, poses a risk that the United States and other Western nations could find themselves at an unfavorable position in the future of the military strategic balance of power.

The reason is that the West possesses two key technologies that China failed to copy and master, no matter how far its economy advances, transforms, and opens wider. One is the technology of the combustion engine, mainly in car powertrains. The other is this cutting-edge semiconductor technology.

Current high-performance semiconductors for military and other applications incorporate various innovative technologies that are shaping the future across manufacturing processes shared worldwide. By way of example, the United States leads in semiconductor design and operational simulation; Japan and the Netherlands dominate manufacturing equipment and materials; Taiwan excels in contract manufacturing; and South Korea is exceptional in the memory sector. Thus, even for a single piece of high-performance semiconductor, diverse technologies are employed at different critical points along the production line; no single country can monopolize them.

For this reason, it has long been said that if China aims to manufacture high-performance semiconductors on its own, it will need to catch up on all these technologies. No matter how much it invests in workforce and capital, it is still considered too ambitious. Concerning the current global semiconductor field, the US, together with its allies in the West, has been taking proactive measures to prevent China from militarily repurposing these technologies. This included restricting exports of cutting-edge semiconductors and marginalizing China from the global internet ecosystem to prevent the theft of these technologies.

The problem, however, is that most of the high-performance semiconductor technology China craves is owned by three of the four nations hindering its maritime expansion—Taiwan, South Korea, and Japan. The Chinese government is desperate to achieve hegemony in the region, also due to the high-performance semiconductor technology.

There is no doubt that the Chinese government wants to invade Taiwan as soon as possible from a political perspective, but the economic factor is also in play. Now, let's take a look at how the Chinese economy stumbles.

7. China's economic gloom mounts

The remarkable economic growth in China during the 2010s was underpinned by the development of the real estate industry, which accounted for approximately 30% of the economy and served as a crucial source of revenue for local governments. However, COVID-19, which started at the end of 2019, marked a turning point, as the previously robust Chinese economy slowed and records slumped. The economic stagnation caused by the pandemic severely impacted the real estate industry in China. Losing such a pillar led to economic growth collapsing. This finally triggered the full-scale collapse of the long-feared, painful Chinese property sector crisis.

The problem lies in the scale of these losses-China's real estate sector started to thrive in 1995, when local governments actively involved in the transactions of the lease rights with private developers. The debt burden snowballed over a short period, expanding in tandem with this growth. As a case in point, the outstanding debt issued by local governments in China through third-sector enterprises is estimated at approximately $9 trillion, most of which is currently in default. Considering that the collapse of Lehman Brothers triggered the 2008 financial crisis, the debt even skyrocketed $600 billion. The total amount of nonperforming loans during Japan's bubble collapse was about $680 billion, so the losses from the crisis in China are on a completely different scale. Never forget that this figure accounts only for the debt held by local governments. The scale will be impossible to imagine if we expand our discussion to include banks' nonperforming loans.

Moreover, an even bigger problem is that the impact of this crisis, compounded by the Chinese government's mismanagement, further worsens the lives of its citizens. The primary causes include the following three factors,

1. Beijing prioritized monetary policies rather than income policies in response to COVID-19.
2. The ineffective legal system of "personal bankruptcy."
3. The polarization of income between urban and rural areas due to the household register system

As an illustration, the "Evergrande Group" is a symbol of the property sector crisis in China. The corporation has been a giant dragon in the Chinese real estate industry for a long time. It was delisted from the Hong Kong stock market on August 25th, 2025, and will proceed directly into liquidation. The actual problem lies in the scale of the losses. Looking at Evergrande Group's total debt alone, it soared to approximately 250 billion RMB, nearly half the total amount of nonperforming loans Japan suffered during the bubble economy. When we take one step further to glimpse the entire market, the figures are even more unimaginable. The massive nonperforming loans generated by the economic bubble burst also directly affect Chinese banks, leading to widespread credit tightening across the banking sector.

At the time the pandemic took hold worldwide, Beijing sought to mitigate the economic impact through monetary easing rather than stimulus (income policies) due to concerns about worsening government finances. Fiscal policy, such as simply distributing money to citizens and businesses or implementing tax cuts, adds to the national debt. However, in monetary policy, if the government merely lends money to struggling individuals and companies, the money will eventually be repaid. This avoids permanent financial damage to the government to a certain extent.

The bubble burst that followed the COVID pandemic left banks saddled with massive amounts of nonperforming loans. At this time, the government's monetary easing policies further worsened the lives of Chinese citizens. This was because banks, struggling financially due to the crisis, had to impose strict debt collection on their clients. Individuals and companies who had loans during the COVID period were also caught up in this.

Under the current BIS regulations (capital adequacy ratio, the "CAR), banks for international transactions are required to maintain an 8% capital adequacy ratio, and even those that only handle domestic transactions must hold 4%. This, in turn, means that banks have a maximum limit on the amount they can lend, determined by the size of their assets. If the value of a bank's assets plummets, say, due to a bubble burst, the bank has to stop all its operations until its client loans are settled.

I have eye-witnessed the never-before-seen bubble burst in Japan, so I understand exactly what banks do during an economic stagnation. They typically don't merely tighten lending; they desperately attempt to sustain their operations by forcibly collecting debts from clients through a practice known as "loan strip," even as their capital adequacy ratio plummets and the economy is pushed into contraction. Targets are not limited to businesses with loan default; banks don't hold back even against trusted corporations with financial successes. But why is that? Because it helps them get out of debt much more easily. So, if the government says yes to a downturn and makes it persistent, it becomes a prolonged recession. This, in turn, worsens banks' capital adequacy ratios, leading to even more relentless collection from their clients. It creates a vicious cycle where the adverse effects spread to ordinary, law-abiding businesses and private citizens who have nothing to do with the bubbles.

Let's not forget that China is a new member of the WTO, and its legal framework for operating a capitalist economy remains inadequate, as does the legal framework for bankruptcy procedures applicable when individuals default on their debts. China joined the WTO (World Trade Organization) in 2001 to engage in formal trade with the world's capitalist economies. It started off pretty well, but things went wrong after that.

One issue was corporate bankruptcies. When China joined the WTO, the Chinese government and courts rarely recognized corporate reorganizations or bankruptcies, so foreign companies were forced to merge with local firms to enter the Chinese market or to build factories. The strategy helps China to catch up with other advanced nations in technological capability. During these mergers, international firms were obligated to transfer their pioneering technologies to their local partners in China. To put it plainly, a proper "Bankruptcy Law" should function for companies that fail in China to retreat. Still, China didn't establish any. It is intended and strategic.

Yet, bankruptcy is a crucial element of the capitalist economic system. The practice of China does not fit in here. It took another six years for China to establish its "Bankruptcy Law"... the so-called "Enterprise Bankruptcy Law". The problem, however, is that the system only covered corporate bankruptcies, with no legal framework for personal (natural person) bankruptcies. This guided Chinese people through tough times in life, especially those who purchased real estate property when the market ballooned, as they could no longer repay their debts in the bubble.

Of course, the Chinese government is wide awake to this problem and is currently rushing to enact bankruptcy laws for individuals. First, on May 18th, 2020, the Central Committee of the Communist Party of China and the State Council issued the "Opinions on Accelerating the Development of a Socialist Market Economy System for a New Era", stipulating that "the bankruptcy system will be improved and the legal system for corporate bankruptcy will be reformed; the legislation of personal bankruptcy will also be promoted to enable entities and individuals in default to exit from the financial market in a system-based manner". Considering this trajectory, we can expect a "National Personal Bankruptcy Law" to be enacted soon.

Nonetheless, the bubble has already burst, and society's festering issues have come to the surface. Until the enactment, a large number of Chinese people will still be forced to live their suffocating lives with massive debt repayment in a tunnel without end. The sudden and dramatic decline in China's birthrate stemmed from these underlying issues, revealing the harsh reality of the living standards in China.

Moreover, this problem extends far beyond mere personal debt, as it now encompasses the debt arising from loans banks extended to corporate entities. This is because most Chinese banks require the owners of companies, such as local startups, to provide joint and several guarantees for corporate debt when lending to these businesses.

When we turn on the TV, we almost always catch a story about Chinese uninhabited "ghost" cities that seem to have been abandoned after years of construction, following the collapse of the Chinese bubble. They mirror a bubbling debt crisis, funded by ballooning debt, that those companies couldn't survive with all they have been dealing with. At the end of the day, even when corporate bankruptcy is legally recognized, struggling companies in business failure cannot liquidate their operations if the owner's personal bankruptcy cannot serve as a joint guarantor. This leaves business owners with no option but to do a moonlight flit. This issue is also one of the major reasons the resolution of nonperforming loans from the Chinese bubble is hardly achieved through credit recovery.

Finally, China's household registration system accelerated income polarization between urban and rural areas, further complicating

China's economic slump. Such a household registration system doesn't make sense to Americans at all. Oriental societies (i.e., China and Japan) are deeply rooted in Confucianism, so it is a tradition for the state to manage citizens' places of origin and family lineage based on registrations. This is referred to as the "hukou system" in China, which places restrictions on where people can live and is monitored by the state. The system made it extremely difficult for Chinese people from low-wage rural areas to move to high-wage cities due to the grading system.

The restriction on the location of dwellings through such a "household registration system" is, in turn, polarizing wealth in China. It also pulls down China's GDP per capita. To put it simply, the current business market in China is far from being unified. It is not a massive market with healthy product flows and desirable mobility. Instead, all provinces in China operate as isolated countries, resulting in rising economic inequality.

When COVID-19 upended our lives, bubble concerns teed up, and a downturn hit real estate prices as demand just shrank. To tackle the issue, Beijing should have abolished the household registration system to boost per capita personal income and GDP, so the burden of total debt can swing down. The stress stemming from a $100,000 debt on someone earning $1,000 a month and someone earning $1 million a month is totally different. Increasing per capita personal income and reducing the burden of debt are thus the keys.

Did Beijing move in this direction? No. No, for several reasons. First, it contradicts the "One Belt, One Road" policy upheld by Xi Jinping as part of China's economic hegemony strategy, as evidenced by the country's status as the world's factory and its extensive product supply. Beijing sought to keep wages as low as possible to attract foreign investors to establish factories.

The second reason is that if the household registration system were abolished to allow Chinese citizens to move freely, to choose where they live, the problem of rural depopulation would accelerate. Such outmigration also accelerates the accumulation of nonperforming loans in those developing regional cities. The construction boom that accompanied China's rapid economic rise sparked a wave of projects across the country, giving rise to vast, under-inhabited developments, including entire pre-built metropolises that never filled up, earning them the now-famous nickname "ghost cities," with the famous examples of Fengdu Ghost City in Chongqing and

Ordos City in the Inner Mongolia Autonomous Region. These vacant housing properties surfaced as demand has shrunk due to factors including rising home affordability, an aging population, and slowing population growth. The more ghost cities there are, the greater the rise in nonperforming loans. And that is why China is so tied up with the household registration system.

Hello, Beijing. Let me make it super clear-the quickest way to minimize damage from cleaning up the bubble is to swiftly write off uncollectible debts as losses and have your office properly replenish the resulting capital adequacy ratio losses for banks. Clinging indefinitely to the "One Belt, One Road" initiative while fixating on unattainable global hegemony will only lead to further damage.

During the COVID pandemic, when China relied solely on monetary policy, its economy deteriorated due to Beijing's reluctance to spend. I couldn't help but see Xi Jinping's economic incompetence mirror Japan's Ministry of Finance. (When Japan's bubble economy collapsed, Japan's Ministry of Finance, concerned about worsening fiscal conditions, said no to capital injection until the very end, ultimately causing the national debt to balloon further.)

In particular, Xi Jinping has purged numerous financial advisors, viewing them as political enemies for the projection and consolidation of his power over China. Growth has vanished under Xi, and China's economy is shrinking to the point that the government is unable to reverse the trend.

Let's not lose sight of the essence of the problem: indicators are now raising alarms about Xi's obligations at the 2027 Communist Party Congress. He is undoubtedly planning to take some action regarding the poor performance, and one of the biggest concerns is the possibility of a military invasion of Taiwan. Xi has been preparing for a long time, and China is on the road.

Beijing has been asserting, over and over, again and again, to its people and the international community that Taiwan is part of Chinese territory, so the invasion is very much justified. Xi might use this battle as a bet to once again earn the Chinese people's respect. Based on the tense China-Taiwan relations.

Not many countries recognize Taiwan as a country and maintain formal diplomatic ties with the island. Even the United States, which is very familiar with the military threat posed by China, has yet to form a formal military alliance with Taiwan, unlike the Philippines, Japan, and South Korea.

In addition, the current dispute between China and Japan over the sovereignty of the Senkaku Islands is just one step in the preparations for the Chinese invasion. If the Chinese army invaded Taiwan, its fleet would need to enter the waters around the Senkaku Islands, which are within Japan's maritime jurisdiction. This is precisely why China has been claiming it owns the islands.

Xi also understands well that his nation will come to realize his responsibilities for the governance failure by the time his current term ends at the Communist Party Congress in 2027. To him, the Chinese military must start invading Taiwan to bring China into war in 2027. We have to rethink the threat that China is very likely to invade Taiwan.

8. World history will change in just 15 years

2025 is the year when power is transferred from Biden to Trump… the economic decoupling between Russia, China, and our Western nations is poised to begin in earnest.

In fact, the Biden administration had, up to this point, taken a tough stance toward China. In business exchanges, however, the separation between the US and China did not go as well as expected. Biden saw victory "when Russia or China suffers economic recession because of the Russia-Ukraine war or China's bubble bursts". He believed that the occurrence would be equal to the trumpet of victory for the West. The reality, yet, did not turn out as he thought. Instead, it developed quite the opposite: the US economy seems stable at a glance, but it was sliding deeper into economic weakness due to aggressive Chinese dumping. The US is wrecked in various ways.

So now, I am pretty sure you will agree on the importance of reassessing China. It is the most potent adversary the US has ever faced. The course of global liberalism and democracy over the coming decades relies entirely on America's actions. Fellows of our lovely modern era remember-life is too short to wake up with regrets.

Taking a closer look at China, I am convinced that the deadliest mistakes in American history were opening the doors of capitalism to a non-democratic country like China. Since the super failure, China has become the world's second-largest economy, and we ought to admit that it is now so close to the USA. Yes, China is experiencing a spell of deflation, or falling prices, that contrasts with inflationary pressures prevailing elsewhere in the world, so it is impossible for it to overtake the US. Maybe. Let's not sit back this early. We'd better embrace the sense of urgency for its military power, as the surpass might come someday. It is a naked truth, and it is something inexcusable. Fellows of our modern era, there's no room for leniency.

It's time to stay alert and make our choice both economically and militarily. In terms of military, I am talking about the Taiwan invasion; in terms of economy, the question is whether we will continue to push electric vehicles merely to eliminate CO_2 emissions, even at the cost of destroying our industries.

First, regarding the response to the first scenario, "if China actually invades Taiwan," the United States and other Western countries need the resolve to defend Taiwan at all costs, even if it means engaging in full-scale war with China.

The key point is whether the superpowers with nuclear weapons are well prepared to start a war. If the defense line in the region comprising Japan, South Korea, Taiwan, and the Philippines were to be breached by the Chinese Navy, China would inevitably expand its massive military power into the Pacific, as a threat to the security of Western nations. So, it is not just the USA, EU members should also get ready. To prevent this, it is absolutely necessary to halt the expansion of Chinese military forces at the stage of a Taiwan contingency, ensuring Beijing cannot go too far.

It's time to make sure we are on the same page-the problem here is that even if the United States and its allies in the West were well-prepared, we are not quite sure whether Japan and South Korea are on board. As mentioned, I lived my life in Japan, and I found that people from both Japan and South Korea are blind to the realities and have no sense of an impending crisis. Should the Taiwan contingency arise, it would be an invasion and a war, yet no one feels the crunch at all.

This is toxic for the US, and this makes no sense. Why do South Korea and Japan have a negative attitude and refuse to cooperate? It is your territory and your issue, but why are you watching the drama over the US and China? You guys are a bit loony at times!

Besides giving craziness back to crazy South Korea and Japan to ward them off for inanity, the US also needs to adopt a macro point of view for better analysis of the issue. South Korea and Japan expressed their opinions in such an unusual way because China scares them.

Let me question this one more time-why didn't China export "fentanyl" directly to the United States, but instead make a detour through Japan, Mexico, and Canada to get it into the US? It's because the true, real purpose was not to weaken the States from within using illegal drugs like "fentanyl," but rather to sever the relationships among these countries to isolate the US.

Let's think it over. If you were Xi Jinping and you wanted to claim Japan and South Korea as your territory, what would you do? Easy breezy, you send a large number of neutralized Chinese citizens into the political, judicial, and administrative spheres of Japan and South Korea, and have them dismantle the military alliances with the United States. Once South Korea and Japan are weakened from the inside, overstepping authority would be as easy as ABC.

Yes. If you were Xi Jinping and you wanted to claim Japan or South Korea as Chinese territory, you might still think that's too ridiculous... but no, it's not. What is happening with South Korean politics right now? Lee Jae-myung, the President, will definitely reduce US forces. And in Japan, where I live, the number of neutralized Chinese nationals who got into politics and courts is increasing every single year, as more intellectuals and journalists in Japan are sounding this same alarm on social media and many other channels of such internal invasion by Chinese nationals.

The Liberal Democratic Party's crushing setback in the June 2025 Tokyo prefectural election was a stark manifestation of Tokyo residents' dissatisfaction with the party's handling of foreign-related issues. In recent years, overt acquisitions of real estate and businesses by Chinese nationals have become increasingly noticeable in major Japanese cities like Tokyo and Osaka. The danger is that the properties involved are not limited to residential condos, commercial properties, or retail shops. They also include areas in Hokkaido known for their pristine water resources, cremation service providers, and many other unusual properties for the weirdest industries. The crushing defeat in the Tokyo prefectural election reveals Tokyo residents' dissatisfaction with the LDP government's failure to act against the Chinese, who are now squeezing into every aspect of life in Japan. At the same time, it also shows how thorough Beijing's preparations are for this Taiwan invasion.

When we think about what the world could look like in the future, the United States and China, the world's two biggest economies, will definitely be facing more and more tensions, with implications for the global economy. They might stand up to each other at any time. So, which side should Japan and South Korea take when that happens? The USA or China? The best and most peaceful way would be to organize a poll and check the will of the people first, but that would be something that happens in Utopia, not in South Korea or Japan, because China will never safeguard such fair play. It stops at nothing in pursuit of its ambition.

Let's say Japan wishes to keep staying with the United States and other Western countries in the international community as it has until now. To maintain its pivotal global role, Japan must provide defensive support once China invades Taiwan. Then, China would definitely speak up rather than shut up, but the problem is: how exactly would China take the stand? It would probably start the stir with Japan's left-wing and use Japan's pacifist constitution as a shield to launch a massive campaign to stand its ground. China is a sly old country. It is horrendous. The strange but actual fact is that both Japanese and Koreans are too gullible to be tricked by the constitution of peace. They have no clue about the serious consequences of a Taiwan contingency that they will both vanish within a decade.

Dealing with China requires a realistic view of its capabilities, not only for military reasons. From an economic perspective, China has become richer, stronger, and an unmistakable threat. It is scheming behind the scenes to seize control of the global economy, taking advantage of international concerns over climate change. It deceitfully promotes electric vehicles as a new transportation alternative. Obviously, China aims to eliminate our life buddy of traditional gas vehicles, using CO_2 emissions as an excuse.

China talks about being eco-friendly?! Who are you kidding? The hidden goal is not our environment but their ambitious hegemony. Yet, is the West that innocent of this scary intention? I am afraid not. They are not finger-pointing at China merely because they cannot control the trend of electric vehicles. Even if they try, China won't let them.

I believe it is feasible to cut off China from the international community and to kick out its fake-nice EV promotion efforts. Our traditional automobile industry deserves to be preserved. Our world has changed significantly since China began developing its electric vehicle project. We now have much more advanced technologies to reduce CO_2 concentrations, as well as many other alternatives in other dimensions. We now have a much broader perspective on protecting the global environment. Just remember, behind the fabulous talks about environmentally friendly EVs is a deceit of China's hegemony.

A decade after the Paris Agreement, we have reached a time to reassess the issue of global warming. Let us stop pushing electric vehicles any further. We are not being friendly to our world by doing so.

Chapter 2

China's coercive economic tactics

1. The background of electric vehicle promotion

The trend toward electric vehicles as the next-generation, zero-emission transportation solution stems from the 2008 financial crisis, the very beginning of the history of "sustainability" and the word "sustainable".

At that time, US President George Bush was desperate to boost his approval ratings for the election for his second term to carry on his high strategy of the Iraq War, which didn't seem to be making any progress. To regain public trust, he implemented a massive economic stimulus plan through a series of aggressive, collaborative monetary policies with Federal Reserve Chairman Alan Greenspan. Unfortunately, that backfired, triggering the well-known American stock market bubble and the collapse of Lehman Brothers.

After Lehman Brothers filed for bankruptcy, the world experienced a financial crisis and a deep recession. The impact on the US was tremendous, with immense losses in the American insurance industry and a collapse in housing prices due to the subprime mortgage crisis. The impact also spread to reach the Big Three-GM, Ford, and Chrysler, which had long been the driving force of the American auto industry.

The development of General Motors, for example, was not limited only to production and sales at that time. A large portion of its profit comes from its finance division, which provides low-interest auto loans to low-income car shoppers, and this has been the key driver of its vehicle sales figures. These loans began to default around 2005. Time to get the red pen out. The Lehman bankruptcy and the US recession wiped out all its business recovery attempts.

This unprecedented financial crisis turned GM's thriving business upside down. It continued to incur huge losses, basically in its financial division, and in 2009, the giant, which once shone through and owned the times, was finally pushed to the brink of bankruptcy. The US government had its back, so it didn't approve the closure of such a colossal corporation's business. It quickly adopted a policy of injecting public funds and, to a certain extent, rejuvenated the corporation.

The real question here is what happened to General Motors' management after that? It showcased a successful change that reinvented the game in just one year. Its debt payoff strategies helped it regain financial freedom, as all government loans were paid off one year after its bankruptcy. Yes, my lovely readers, you guessed it right.

Table 1 General Motors' Financial Releases

	2002	2003	2004	2005	2006	2007	2008
Net Sales (Billion US$)	1,867.6	1,855.2	1,935.2	1,946.6	2,033.7	1,799.8	1,489.8
Operating Margin	19.8	28.6	11.9	-178.1	-21.6	-426.9	-310.5
Net Profit	17.4	38.2	28.0	-104.2	-20.3	-385.4	-309.4
Assets	3,690.5	4,485.1	4,796.0	4,741.6	1,863.0	1,488.8	1,894.8
Cash Flow	203.2	325.5	359.9	151.9	237.7	245.5	151.5
Total Debt	3,622.4	4,232.4	4,518.8	4,584.6	1,919.6	1,859.8	1,676.2
Interest-Bearing Debt	2,001.7	2,717.6	3,002.8	2,877.2	425.1	443.4	1,263.2
Ownership Equity	68.1	252.7	277.3	146.5	-56.5	-370.9	218.5
Ownership Equity Portion (%)	1.8	5.6	5.8	3.1	-3.0	-24.9	11.5
Total Sales Volume (10,000 units)	841.0	810.0	824.0	838.0	909.0	936.0	835.6
Sales Volume in China		20.1	49.2		87.7	103.2	109.5
President	**George W. Bush**						
Major Event		Iraq War		Expanding low interest rates on car loans			Bankruptcy of Lehman Brothers

Main Source- POSITEN (https://positen.jp/806)

2009	2010		2015		2020	2021	2022	2023	2024
1,045.9	1,355.9		1,523.6		1,224.9	1,270.0	1,567.4	1,718.4	1,874.4
-210.2	51.1		55.4		66.3	93.2	103.1	93.0	127.8
1,046.9	46.7		96.9		62.5	98.4	89.2	100.2	71.9
1,723.1	1,389.0		1,943.4		2,351.9	2,447.2	2,640.4	2,730.6	2,797.6
147.9	278.6		234.0		290.4	286.8	313.0	264.7	271.4
1,514.7	1,017.4		1,540.2		1,855.2	1,789.0	1,921.1	2,048.8	2,141.7
983.1	104.7		435.5		729.8	756.6	759.2	827.7	903.0
208.4	371.6		403.2		496.8	658.2	719.3	681.9	655.9
12.1	26.7		20.7		21.1	26.9	27.2	25.0	23.4
748.0	838.0		984.1		682.9	629.1	593.9	618.8	600.0
182.6	235.0		373.0		290.0	289.0	230.0	210.0	180.0
Barack Obama				**Donald Trump**		**Joe Biden**			
Business failure (public fund injection)	Public funds repaid & Euro area crisis		Paris Agreement		COVID19 & Brexit		Russo-Ukrainian war	Israel, Gaza War	

The wizard wand behind this miracle was China. At that time, the Chinese economy was booming, achieving double-digit growth year after year (unlike the current, fast-declining trend). At that time, no one had a knack for choosing a business partner. Corporations kept exporting cars to China, including General Motors. The strategy was robust, and the American multinational automotive manufacturing company soared.

General Motors was not the only one to be fortunate. On the same roller coaster run by the Lehman bankruptcy were Ford and Chrysler. They also found success exporting cars to China and achieved a V-shape recovery. (The case of Chrysler is somewhat more complicated, as FCA acquired it during its financial crisis, so the credit should also go to FCA.)

"China is right for my business! We are raking in the cash!"

No, no way. As I have illustrated multiple times. Doing business or investing in China looks great at first, but you will end up burning through money.

Table 2 Annual Sales of Major Global Automakers in China

	2015	2016	2017	2018	2019	2020	2021	2022	2023	2024
GM (General Motors)	373.0	387.1	404.0	355.4	309.0	290.0	289.0	230.0	210.0	180.0
Ford	111.5	127.0	120.0	75.0	53.0	61.0	64.9	49.5	46.7	44.2
Tesla						13.8	30.7	44.0	60.0	65.7
Volkswagen	257.7	299.7	311.4	301.8	314.3	262.0	242.8	239.7	323.6	292.8
Toyota	112.0	121.0	129.0	147.0	169.0	179.0	192.0	187.0	190.2	178.9

Main Source- POSITEN (https://positen.jp/806) Unit- 10,000 units

The table above shows the annual sales trends of major global automakers in China, where we can see that vehicle sales by the American Big Three have steadily declined every year since 2017, except for Tesla. The fundamental cause was the emergence of Trump's first administration and the escalation of trade disputes with China. It was, however, just a trigger. Beijing had been gradually expelling foreign automakers from the Chinese market, making environmental concerns an excuse to grow its own EV industry.

2. China's automotive strategy —taking advantage of the Paris agreement

Here, let me make the long story of global warming short. The total amount of carbon (the element) on Earth has remained constant since ancient times, i.e., the quantity has not changed. Yet, it has existed in different forms. Have you ever heard of carbon in solid form? What is it? Any good guesses?

Is it diamond or graphite? Well, they are made up from carbon, but the answer is "plants", the familiar existence in our daily life. Plants that fossilize in solid form become "coal," while those that fossilize as liquid become "petroleum," but the components in both are almost all carbon. The problem is that the form of carbon, whether it is solid, liquid, or gas, has an entirely different impact on the Earth, our lovely habitat. A top current global issue we must address is that when carbon exists as a gas, that is, when it is in the form of carbon dioxide, it insulates our planet. As CO_2 levels rise, global temperatures continue to soar, driving a pressing crisis that signals the onset of mass extinction.

Then, why are carbon emissions increasing? One major contributor is the ever-growing energy demand. Countries are racing to improve their infrastructure and living standards, often relying heavily on fossil fuels such as coal, oil, and natural gas to power their economies. Burning fossil fuels, such as petroleum, is indeed the most energy-efficient method. Better examples are thermal power, the natural gas engine in automobiles. The more human society pursues economic growth overall, the more energy it consumes. Coal and oil consumption in this process are released as gases (CO_2), causing more heatwaves and climate change events to become frequent and more intense. This is why global temperatures are spiking.

To address these issues, the 21st annual session of the United Nations Climate Change Conference was then held in Paris, France, in 2015. At this conference, the goal of achieving net-zero emissions by 2050 was created to translate related goals into a global pathway for the energy sector (commonly known as the Paris Agreement).

Achieving net-zero emissions by 2050 means, of course, we must unleash global energy to transition away from oil, coal, gas, and all other similar sources. The CO_2 profile of Japan, my country of origin, is 30% from electricity generation, 30% from transportation (e.g., automobiles), 30% from corporate emissions, and 10% from other sources. To curb emissions, automobiles are the key. And, is ending the sale of gas cars the way out, or is it a radical idea?

The problem is that the world struggles to stay focused on this grand vision. China longs to become a hegemon, so it steals our dreams.

2. Global EV outlook

In the 2024 United States presidential election, which happens every four years, Republican Donald Trump made a historic return to the White House. He swept to a decisive victory as Americans gave him a crystal-clear mandate for change. He also won so big because of the backfire stemming from the political failure of the "EV promotion policy" proposed by the former Democratic President Biden.

To begin the story, it was China that chose electric vehicles for its next-generation cars. In recent years, it has evolved from a poor, backward nation into the world's second-largest economy, achieving numerous technological milestones along the way. Despite its economy emerging stronger with new growth, one technology that is said to remain far lagging the West is the internal combustion engine for automobiles and its associated drive systems. This field sets the tone for Beijing's ambition that China has stopped fancy innovations in combustion engine technologies. Instead, it shifted to its next-generation transportation to electric vehicles with an apparent aim to stave off the worst consequences of climate change. This is, in fact, calculated much more based on geography and practical challenges.

As captioned in the previous chapter, China has some distinct geographical features. Its littoral zones are microscopic compared to its large, hard-to-envisage size, so the country has developed into an economic, military, technological, and diplomatic power, yet has no active nuclear energy policy. This is because operating atomic power plants requires vast quantities of cooling water for the reactor core. China doesn't have any coastal plants, and it finds it difficult to create inland ones. In case of a nuclear disaster, radiation leakage can dissolve in rainwater and flow directly into the sea only if it occurs along the coast. If a radioactive leak occurs at an inland plant, on the other hand, it will remain trapped, contaminating the entire region and making the area uninhabitable. Having no next for nuclear power, China needed an alternative.

Since the sunshine duration in northern Chia, a massive desert area, is long, renewable energy sources such as solar and wind power can be developed as alternatives to nuclear power, as a wise land-use decision. The drawback is that these power generation methods depend heavily on weather conditions, so that electricity output can vary significantly. As a result, large batteries are crucial for storing purposes. It is exactly China's development strategy, and the progress of these technologies has, in turn, led to widespread EVs.

As you might already know, it's often said that the core technology behind EVs is all about the battery. For EVs to achieve driving ranges comparable to those of traditional gas cars, there is still room for improvement. Most of these issues stem from the battery itself, including charging time, efficiency, miniaturization, and safety in a crash. As mentioned earlier, however, China has been watching battery technologies, energy storage, and related insights as a fundamental national issue, so they are also applicable to electric cars. As China dominated the automotive battery market with a share of over 50%, Beijing has no reason to focus on gas cars and go through the struggles of the combustion engine. The shift to a "next-generation transportation method" powered by electricity is the one and only option for China.

Down to earth, it was automakers in the West who, by embracing the Chinese government's global push for EVs, sealed the deal for EVs to become our future. In recent years, the evolution of China has spurred significant expansion in automotive exports worldwide, and the Chinese market has become one of the world's largest. This saved America's Big Three from the worst operational crisis with desirable by-products because the Chinese market had been so profitable for Western investors. The shift in automotive sales toward EVs in China has already become an unstoppable trend, though no similar signs are evident in Western markets. China is, after all, too ready to sell EVs.

In addition to substantial subsidies for electric vehicle promotion, Beijing implemented a series of measures targeting gas cars in 2009, including restrictions on license plate issuance and limits on driving days, apparently to address environmental concerns and traffic congestion resulting from inadequate transportation infrastructure. Quite the opposite, electric vehicles are free from all limitations, so the shift to this new form of automobile progressed almost overnight in China. Today, China has risen to become the world's second-largest EV market by penetration rate, trailing only the EU.

Executives in the automotive industry in Western countries had been grappling with the idea that "solving the global warming problem would require enormous ongoing investment to curb vehicle emissions." As EVs are now selected to lead the next generation, the targets for investment have also been clearly defined. Improved charging speeds, lighter vehicles, development of all-solid-state battery technology to reduce the risk of battery fires, driver assistance technologies to enhance safety. Thanks to China's decision to go all-in on EVs, our path is clear. Investment funding? All set. The next step is to witness how EVs take over the growing Chinese market by storm.

Big automobile promotion in China's market and the complete shift to EVs with zero emissions by 2050. By that time, EV manufacturers will have monopolized the market, and gas car makers will be marginalized. All car companies believed this before the coronavirus outbreak, except for Toyota. EVs are indeed the key. It is an innovative means of transportation for the next generation.

At that time, Biden followed the trend and implemented several policies in his administration, including substantial subsidies to promote EV adoption at the cost of national deficits. Still, this cast a significant shadow over the American auto industry. Why?

Switching from gas cars to EVs meant abandoning gas-engine technologies and the accumulated efforts behind them. Enterprises had to start over at zero.

In simpler terms, China scored when our guard was down during Biden's presidency, when the most rigorous conditions were created.

4. Joe Biden's failed presidency

China was the first to step forward to pick EVs that emit no CO_2. Yet, it was the USA and the EU who crafted the global promotional plan.

The ideal picture for the EV transition should have been a step-by-step shift in the propulsion system of a single vehicle, starting with gas cars, then hybrid cars combining a conventional engine with an electric motor, and finally battery electric vehicles (BEVs). The role of the motor is becoming increasingly prominent along this path. This is because the worldwide annual production capacity for automotive batteries was still far from sufficient at the time.

Switching all vehicles to EVs was thus a snowball's chance in hell. It was considered far more effective for curbing emissions to allocate the limited battery supply in small batches to hybrid vehicles first, before any full-scale expansion.

Biden, however, was in the dark when he planned his policies. Instead of introducing hybrid vehicles as an intermediate step, he pushed electric vehicles directly into the market, with an aggressive goal of all EVs by 2050. The head-in-the-sand approach was to compete with Japanese cars. Biden thought he would have made America great running ahead of Japan by this sudden jump, but he had forgotten that China was watching. That was the most terrible mistake Biden ever made throughout his life.

Plus, the US government has chronically made awful mistakes in its automotive policy. One of these was its concern about China's theft of high-performance semiconductor technology. Japan also reflected this concern in its strict export controls and even Internet disconnections with Western countries. Since EVs run on electricity only, they are, in fact, giant mobile home appliances. They heavily rely on semiconductors because of their structure. But the question is, what kind of EVs will the future automotive industry need and create? This should not be those cheap cars that Chinese manufacturers are so keen to sell, but something with more added value. China has already captured most of the global market for automotive battery materials, a key component of EVs, and, in a way, no European country can compete with Chinese manufacturers on price, even if they want to keep pioneering in the industry.

The expected "future image" of EVs was that they would become a safe and comfortable means of transportation to reduce traffic congestion, thanks to advances in autonomous driving technology. To achieve this, however, it was necessary to equip EVs with high-performance semiconductors and to create a fast network environment, such as 5G, to handle the large amounts of bidirectional data transfer for remote control.

Unfortunately, due to Biden's mistaken "China Containment Policy", both technologies could no longer be included in EVs intended for export to China. In other words, not being able to export high-performance semiconductors to China also meant that well-equipped automobiles could not be exported there either. As a result, automakers around the world who had been longing for this expansion were completely out of place. Their finances were off.

Joe Biden's big blunders about China tarnished his legacy. He planned to celebrate when the former Eastern Bloc countries would suffer from significant financial anxiety and exhaustion from the Russia-Ukraine war, and China faced an economic slump, so Western countries would, on the contrary, gain economic dominance. It's unfortunate, but that didn't happen. Instead, the States was at a high level of indebtedness, inflation had soared, and the American dream had grown increasingly out of reach for millions of people. The prolonged slowdown in the Chinese economy led to a firm dollar in the foreign exchange market and sluggish EV exports to China. At the same time, the US was hit hard by Chinese manufacturers' export offensive, which was very much like a dumping policy, so the automakers around the world that were making a significant shift toward EVs at the time bore the most tremendous economic burden. The threat was not only to the auto market. Manufacturing in the USA, including steel and shipbuilding, had been pushed to the brink of collapse amid fierce competition with China and a rapid rise in the USD.

Once EV sales fell far short of expectations, layoffs hit the US. To make matters worse, astonishing price hikes were in store when unemployment rates are high. This ultimately led to further struggles of democrats, so Donald Trump regained voters' trust. Soon after his return, he has been taking swift actions to accelerate EV growth, including directing the US withdrawal from the Paris Agreement.

"EVs are not getting reliable, and they still lag behind gas cars. They don't sell at all. If we still abided by the Paris Agreement, all American auto corporations would go bankrupt!!"

This is what Donald Trump declared and stood on.

5. EV reliability Issues

Setting aside the Biden administration's worst policy and management failures, we also need to consider whether EVs are really a good choice. As mentioned, we have debunked myths.

Carbon neutrality will be achieved in 2050... we used to believe that we are determined racers on the field. We used to trust this key strategy to curb rising global temperatures. Automobiles with no emissions, environmentally friendly electricity generated by solar and wind power in EVs and households... thus a sustainable, ideal society. And yes, we will witness the harmonious coexistence of humankind and nature… someday, maybe. Is it true?

The mining, refining, and transportation of rare metals used in EV batteries, such as lithium, cobalt, and nickel, produce significant emissions that degrade the environment. EVs can only reduce emissions while the vehicle is running. Manufacturing? A different story. Furthermore, there have been many incidents of inhumane exploitation, such as child labor, occurring blatantly at rare earth mining sites and other resource locations. These are now what we recognize as a global issue.

Annual CO_2 Emissions by Major Countries (Billion Tons)

	2001	2002	2003	2004	2005	2006	2007	2008	2009	2010
USA	57.5	55.9	56.6	57.4	57.5	56.5	57.4	55.6	51.6	53.9
China	35.3	38.0	44.2	51.1	58.2	64.3	69.9	72.0	77.2	84.7
Japan	12.5	12.8	12.9	12.9	12.9	12.7	13.1	12.4	11.7	12.2
UK	5.5	5.3	5.4	5.4	5.4	5.4	5.3	5.2	4.7	4.8
Germany	8.5	8.3	8.4	8.2	8.0	8.1	7.8	7.9	7.3	7.7
France	3.8	3.7	3.8	3.8	3.8	3.7	3.6	3.6	3.4	3.5
Italy	4.4	4.4	4.6	4.7	4.7	4.7	4.6	4.4	4.0	4.1
Canada	5.1	5.2	5.4	5.4	5.5	5.4	5.7	5.5	5.2	5.4
India	9.5	9.9	10.1	10.9	11.4	12.2	13.4	14.2	15.6	16.6
South Korea	4.6	4.6	4.6	4.9	4.8	4.9	5.0	5.1	5.3	5.8

Source- World Economy Notebook (https://ecodb.net/)

An even more fatal problem with EVs is that recycling solar panels is more complicated than you might think. In Japan, where I live, the industry is struggling because solar panel components contain highly toxic substances, such as arsenic, which makes recycling them a big obstacle. It is, in fact, much cheaper to landfill them, so more and more agencies have shifted to landfilling to avoid the costs. In Japan, quite a large number of solar panels are expected to reach the end of their lifespan after 2030. Without a breakthrough, such as thorough legal reform, soil contamination could have a petrifying impact soon due to the unstoppable trend of disposal. Isn't it greenwashing? Are the moves truly sustainable?

2011	2012	2013	2014	2015	2016	2017	2018	2019	2020	2021
51.7	49.6	50.9	51.1	49.9	48.9	48.2	49.8	48.2	43.2	43.2
92.8	95.4	99.8	100.2	98.6	98.6	100.9	105.7	107.6	109.4	
12.7	13.1	13.2	12.7	12.3	12.1	11.9	11.4	11.1	10.4	10.6
4.5	4.7	4.5	4.2	4.0	3.8	3.7	3.6	3.5	3.1	
7.5	7.6	7.8	7.4	7.4	7.5	7.3	7.1	6.6	6.0	
3.4	3.4	3.4	3.1	3.1	3.1	3.2	3.1	3.0	2.7	
4.0	3.8	3.5	3.3	3.4	3.3	3.3	3.2	3.2	2.8	
5.5	5.5	5.6	5.6	5.6	5.6	5.7	5.8	5.7	5.2	
17.6	19.1	19.7	21.5	21.6	22.0	23.1	24.6	24.2	22.0	
6.0	6.0	6.0	5.9	6.1	6.2	6.3	6.3	6.1	5.7	

In solar power stations, the heat accumulated on their large solar panels can also cause abnormal temperature rises in the surrounding areas. As for wind power plants, the loud noise from wind turbines is also a problem for the local communities. Here is where good intentions meet harsh reality. These sustainability claims sparked backlashes, eroded trust, and triggered one problem after another. In the end, renewable energy sources such as solar and wind power can only be installed in remote areas, like deserts or oceans. The larger the country's population, the more serious the installation and operational problems.

Another problem is that the number of these "eco-friendly" power stations will soon lag behind the demand from electric cars. Which corner of the island country called Japan can accommodate more renewable energy installations? What about larger countries like the USA and China? Are they really that big for the large factories that might be needed in the future? Let me tell you the truth: the onus of achieving carbon neutrality does not lie solely on the USA and China. Global warming is a worldwide issue that requires the involvement of all countries. Otherwise, it can't be stopped.

Bad news is everywhere. More and more studies warn that the world is running out of time, and neither EVs nor renewable energies are helping us to avoid the worst impacts. Instead, they are just creating more problems in the world, warming at the fastest rate since records began. At the end of the day, the promotion of renewable energy and EVs remains an ambitious plan. They are created to solve problems, but they produce more.

Table 3 (reference 6), "Annual Emissions of Major Countries," clearly shows that despite widespread adoption of EVs and renewable energy in China, the "climate action" stalls. They do not ramp up. The massive ratcheting up of new approaches failed to take meaningful steps. Instead of being phased out, the traditional methods deserve more time and effort so that improvement can be made, so our world won't heat up, and impacts won't get more serious.

The critical warming threshold will be breached as EVs and renewable energies are not helpful. Down to earth, no matter what, China is the only beneficiary-no other countries and companies, not even a single one. Beijing knew it. So, they have been preparing for the acceleration of the "EV-only world" by securing resources, such as rare earths used in EV batteries.

Tabele 4 The Global Rare-Earth Elements (REE) Share of China

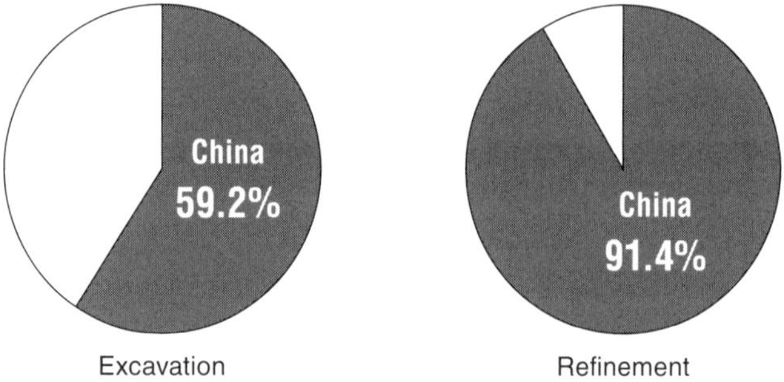

Rare-earth elements	Excavation	Refinement
Graphite	85.5%	95.7%
Cobalt	2.6%	78.4%
Lithium	22.4%	70.2%
Nickel	3.4%	31.3%

Source- International Energy Agency (IEA)

So, when Donald Trump returned to the White House in 2025, he attempted to contain China's state-led economy through tariff trackers in response to China's hegemonic aspirations. However, as we all know, US moves failed to "contain" China; they raised tensions, lowered economic well-being and global economic efficiency, and divided the world economy since China had already come to dominate the global supply of rare earths needed for batteries and automotive motors. The China containment plan failed, and the USD soared, worsening matters. China then even threatened to impose export restrictions on these resources. No negotiation helped.

When Donald Trump won the US election again in 2025, the automobile industry expressed concerns about China's stance on those issues. Japan's Toyota Motor Corporation was one of the companies that remained long skeptical of EVs from the beginning. The automaker was never convinced of the need for a global effort to transition to batteries, citing concerns about a stable supply. Instead, it has been focusing on "hydrogen vehicles" instead. Toyota's candidate for next-generation vehicles is not, by any means, EVs.

Hydrogen cars are fuel cell electric vehicles (FCEVs) that generate electricity through a chemical reaction between hydrogen and oxygen to power a motor. The biggest advantage is that they use compressed hydrogen, which is up to 700 times as fuel-efficient, so, in theory, no batteries are needed.

Toyota is also considering using nickel-metal hydride batteries to explore electric driving, so rare earths are not needed at all. Current nickel-metal hydride batteries are primarily composed of nickel and hydrogen, as the name suggests, so they use almost no scarce resources, such as rare earth elements. Another benefit is that, unlike lithium-ion batteries, which are the primary batteries used in EVs, they emit almost no harmful substances during recycling, which helps keep costs low.

Considering this, it is not difficult to picture why some automakers have long been thinking about EVs wrongly. They are smart enough to know that doing business with China would only bring them short-term benefits. In the long run, they will suffer a terrible loss.

My conclusion is that there is nothing good to come from promoting EVs. It failed as a climate solution. It only makes China the pinnacle of achievement. It is an ugly trend that is bringing our economy to the brink of recession.

So, is "stop promoting EVs" the sustainable solution to solve global warming? Well, excuse me if I led you to stew in awkward silence-we are no longer living in the world in which the Paris Agreement was signed. That is more than a decade ago! Technology for reducing CO_2 concentrations has advanced significantly, and reducing emissions alone is no longer the only way to prevent climate change.

We have only a quarter-century left before achieving carbon neutrality. It's time to pause and calmly reconsider whether we should continue believing in EVs.

Chapter 3

The way EV dies

1. The biggest blunders of global warming policies

The worst mistake about carbon neutrality is that, although we all need to curb climate change by reducing the amount of CO_2 (carbon) in Earth's atmosphere, the means to achieve this have so far been limited to reducing emissions.

It is true that, in a sense, it was inevitable that we had to make such a choice 10 years ago at the time of the Paris Agreement. And yes, that's why electric vehicles and renewable energy became a worldwide trend before we knew it.

Unlike back then, our world has become more civilized, and technology for directly capturing carbon from the atmosphere (DAC) has been growing fast. Our world has reached the point where it is technically possible to control atmospheric carbon concentrations without relying heavily on the troublemakers-EVs and renewable energy sources.

Of course, this needs further improvements in fuel efficiency while retaining current technologies such as gas cars. However, this would be far more cost-effective than the conventional approach of switching entirely to EVs and renewable energy sources and would enable us to achieve a decarbonized society as a low-effort habit. Japan has been focusing on hybrid cars and has been far more effective than the United States, China, and the EU in reducing emissions, even though the latter three have been promoting EVs hard. (Table 3, reference 6, may give the impression that Japan couldn't find access in reducing emissions, but this was due to the simultaneous shutdown of nuclear power plants following the Great East Japan Earthquake in 2011; emissions from transportation have been steadily decreasing.)

But then why hasn't this been touched on until now?

2. The rapidly advancing carbon capture technology

As of yet, reducing greenhouse gas emissions through DOE (direct air capture) has not been considered a true core technology for solving the problem, mainly because of its high costs.

The technology of capturing, utilizing, and storing CO_2 from the atmosphere, known as "CCUS", has been limited to capturing CO_2 released by factories up to now. Usually, the atmosphere contains about 0.04% CO_2. In comparison, exhaust gases emitted from factories contain about 5–20%. The portion is much higher than normal air.

The major obstacle with this technology, however, is that the air must first be heated to nearly 100 degrees Celsius to separate and capture CO_2. So now, let's take a look at a random household appliance nearby-a microwave, an A/C, an electric kettle, or a hair dryer... The higher the heat they give off, the more power they consume. That is, heating the air to 100 degrees also consumes an enormous amount of electricity, which is a destructive or deviant move that harms our environment. We capture CO_2, but we release more. We have our priorities wrong. Capturing CO_2 from low atmospheric concentrations is not cost-effective because the approach works best at higher CO_2 concentrations, such as those in industrial gases. CCUS is out of its depth; it is not up to this kind of eco-friendly grand vision.

Change feels hard, but we ought to move on because times have accepted changes. Some corporations are now working on technologies to capture carbon from the atmosphere at room temperature. One well-known example is AGC from Japan, which has developed a chemical absorption liquid that can separate and capture carbon from the atmosphere at room temperature (25℃). This has enabled reducing energy consumption to approximately 20% when compared with conventional approaches. Needless to say, AGC is not the only one; many other corporations are also working in this direction. So, if future cost reductions are expected, it will be feasible to capture carbon dioxide not only from industrial-sector emissions but also from the atmosphere.

However, the problems with CCUS do not end there. Another issue that remains is what to do with the captured carbon.

3. New ways to utilize recycled carbon dioxide

Until now, capturing CO_2 from the atmosphere has mainly involved "CCU" ... that is, capturing and storing; the subsequent disposal of the captured gas has been primarily and merely keeping it in selected locations deep underground or injecting it into old oil fields where crude oil remains to increase the crude oil recovery rate. These kinds of disposal, nonetheless, all come with problems. Breaking down the real costs of the capture, transportation, and injection of carbon dioxide... the fundamental drivers behind are all expensive. And there are safety issues arising from this, such as leaks and marine pollution.

For instance, in a country with a small total land area, such as Japan, the cycle begins with the most challenging task of locating a suitable storage site. Some countries might even get stuck before starting because they have no potential storage location. Even if we were lucky enough to keep captured CO_2 under lock and key under the sea, the facility construction costs can be scary. Let's not forget the deepening affordability crisis stemming from 24/7 monitoring and leakage prevention expenses. On the bucket list, we also have preventative measures to avoid the devastating impact on marine life in the event of a leak.

4. Chemical fuels that make EV unnecessary

One of the technologies I am focusing on is chemical fuels, such as "e-fuel". They are produced by synthesizing captured carbon

synthetic fuel, and sell it at the same price as gasoline. No one wants to complain, as nothing could be better than such a streamlined strategy to curb greenhouse gas emissions. The effectiveness of EVs is unclear, but this approach is promising-why can't they see? And most importantly, even if they continue to develop EVs, sales to China are not guaranteed. Not at all.

To boot, if artificial fuels such as "e-fuel" allow gas cars to nail it in the future, all the concerns about EVs will be resolved. Think of the time, money, and effort saved from the development of the tremendous infrastructure for EVs. The powerful benefit ensures that countries with a poor government budget (such as the USA) no longer need to raise taxes for environmental protection.

Beyond that, the shift in direction can save many fast-declining industries. Gas stations and auto repairs are perhaps the best examples. Salvation is not limited to the vanishing sectors but also to the auto manufacturers themselves, because EVs generally have around 30% fewer components than gas cars. It has been widely believed that the spread of EVs will lead to mass unemployment in the auto industry. Again, if gas cars can keep running, these workers can keep working.

and hydrogen, like, simply put, artificial gasoline. If we can make it at low cost in the future, we could capture the emissions from gas cars using CCUS and convert them into gasoline for transport. This would circulate carbon dioxide between the atmosphere and vehicles, in turn reducing their emissions.

The benefit of this technology is that the carbon and hydrogen resources needed to refine artificial gasoline, unlike the rare earths used in car batteries, are readily abundant. Therefore, we will be free from the worries about procurement issues such as currency fluctuations and trade friction if we enjoy a stable supply, no matter how advanced the production technologies are. Another great merit is that it will allow all gas cars on the road to continue to run even after 2050.

I personally believe that America's Big Three would have a much more promising future if they stopped developing unsellable EVs and instead focused on developing artificial chemical fuels like "e-fuel" to meet market demand at the lowest possible cost. They could grab concrete figures on how much gasoline each car they've sold consumes each year. They control their vehicles' emissions, capture the same amount of carbon, produce the same amount of

Looking at the progress of these environmental technologies, we may already be at a point where we need to free ourselves from old concepts and adopt a new perspective to strengthen flexibility for our future. Automobiles have been emitting CO_2 into the atmosphere as “waste”, a convenience product we humans must pay for. However, humanity has advanced once more, and the day has come when we should see carbon dioxide as a valuable resource that sustains our lives. It is not merely unwanted waste.

The problem with e-fuel production is that it requires a large amount of electricity and advanced technology to extract, refine, and process carbon dioxide from the atmosphere. Still, the time will surely come when extraction and processing are carried out independently, and carbon dioxide captured from the atmosphere will become available for free trade on the global market. Once such a carbon cycle is created, emissions will become a powerful new force for nature and the human future. In terms of environmental issues, this “invisible hand” also helps and serves our interests.

Witnessing the mitigation efforts and technological advancements as decarbonization initiatives across countries and companies, I can assert with absolute confidence that “even if EVs and renewable energy don't exist, alternatives already do!" In the end, the reason our world has become so obsessed with electric cars and renewable energy is simply that many cannot give up the fantasy of selling cars to China.

So, what is the painfully honest reason we continue to attract this toxic EV trend? Why do we fall for a toxic relationship where all the benefits go to China? We can rewrite our life narrative-selling cars to China made money in the past, but not in the future.

Anyway, the leaders of global automakers should already know that EVs aren't what we need for decarbonization. But what if, by some chance, the automobile landscape has become all about EVs? Well, let's confront the reasons why we were drawn, time after time and until now, for something knowingly bad.

We are now navigating this brave new world. The great shift toward the dead end of EVs might soon be accelerated by Donald Trump's return with a big agenda, a mandate, and the high risk of war in the Taiwan Strait.

5. President Trump's China strategy

Donald Trump was the first US president in history to point out the China threat in public during his first term, sparking intensifying Sino-US trade friction that later developed into an economic trade war and led to a sharp decline in US automobile exports to China. He learned the hard way and grew as a president, becoming much more cautious in dealing with China's authoritarianism and arrogance during his second term.

One of the very first things Donald Trump did after swearing in as the 27th president of the United States in 2025 was to "end the electric vehicle mandate". The actions included revoking Biden-era statements such as "the USA will be the most influential sustainable leaders on the planet". They led to a series of reviews of "EV policies" implemented by Democratic administrations nationwide. Trump then said no state would be allowed to ban gas cars, signed measures to block unfair bans on new gas cars, and vowed to protect the gasoline-powered traditions.

However, his true intention was not merely that he knew that EVs would be of no use in revitalizing American industry or in stopping global warming, but that he fully understood that China would be in trouble if he could introduce a no-EV world.

Take the reciprocal tariffs announced in April 2025 as an example. Its target was the Chinese economy, no doubt. Yet, as I have written a thousand times, tariff increases would never be a winning strategy to contain or defeat China because China already possesses most of the technologies and resources necessary to slow global warming, such as electric vehicles and renewable energy. China and Western countries can no longer decouple easily.

Therefore, the Trump administration has already started planning to edge out China as a threatening competitor in sectors where it has already dominated (such as EVs) and to switch to other sustainable solutions to address global warming. Beijing's "EV tactics" are no joke because electric cars are, in fact, the very heart of China's approach to global governance, named "One Belt, One Road". The Chinese government has used every possible means and spent every available expense to push EVs. The Communist Party is at a point of no return, so afraid that its country will become the only one with EVs in the world by 2050.

If the electric car era doesn't come by 2050, the damage to the Chinese economy will be immeasurable. First, its battery industry and steelmakers will be under direct fire, and Beijing will no longer be able to keep a close eye on other countries for the rare earths used in batteries and motors. The end of China as a great power in battery production will also severely damage its exports of solar panels and wind power generation equipment. If the global electric vehicle industry plunges in this way, China's algorithm will collapse. This policy failure will bring all its ambitions to a halt.

Trump does not lack wit. He knows it well, and that's why he took the EV mandate as his first step in his China containment plan. Trump is not the solo player on the field. The EU has the highest EV deployment rate in the world, but it also sided with the US to stop any further spread due to the slow sales.

The EU's "European Climate Law" stipulates that member states must gradually reduce emissions to achieve carbon neutrality by 2050. Under this, member states must achieve a 55% reduction in emissions by 2030 and a 90% reduction by 2040, based on 1990 emissions. It has become the mission to catch a unicorn, since EV sales are slow in the EU as well.

The EU is now too dependent on China for battery supplies, making it impossible to keep EV prices down relative to gas cars, and leading to slow sales: doom and gloom. Some countries, such as Germany, have ended subsidies to promote EVs due to financial constraints, exacerbating an EV business slump. In response, the EU has begun considering various policies and initiatives in recent years to ensure a smooth transition to decarbonization without causing financial and social impacts. Still, some look on EVs and renewable energy with awe and dread.

The question is, can we stop EVs and renewable energy now? Any shoving that genie back in the bottle? Our world has changed decisively to embrace EVs under Biden's administration and EU policies. How easily can we halt the EV production system verifiably and reliably?

Consumers have already believed them and paid (overpaid), and automakers around the world have poured billions into EV development. From the perspective of private business management, they would not know what to put in their communication letter to their shareholders if they were to stop producing EVs. It's almost impossible to hit pause.

EVs may prove to be a grave threat to the world, but there is a way to fight back and turn the situation upside down once the Taiwan invasion occurs.

6. The national people's congress in 2028

In China, the highest legislative body and the largest assembly that makes national-level policies is the National People's Congress, also known as the NPC. This assembly is held in Beijing, China, every March. The NPC in 2028 will differ from regular ones and be of great significance, as it will mark the end of President Xi Jinping's third term. Whether Xi will be able to continue in office for a fourth term depends on whether he can get nominated at the Chinese Communist Party Congress, which will be held the year before, in the fall of 2027.

Undoubtedly, if Xi is to remain in office for the next term, his current administration will be put up for evaluation. So, what is Xi's achievement? The Chinese economy has been heading toward a recession amid enormous uncertainties, with only one certainty: it will remain fragile at the time of the Communist Party Congress in 2027. Beijing has no solution handy for the economy in this very dire state.

China today not only holds global shares in automotive batteries and their materials, solar panels, but also leads the world in various sectors of the global economy, including steel, aluminum, shipbuilding, AI, supercomputers, communications equipment, home appliances, power transmission facilities, and even the clothes we wear. The mystery of capitalism is that, yet, no matter how competitive or technologically advanced a country might be, this does not turn into the keystone of modern prosperity... not at all in our turbulent century. In other words, even though China has risen to become the world's second-largest economic power, Chinese people are struggling to make ends meet.

Key positions in branches of the Chinese government are indeed filled with Xi Jinping's backers. Xi's power base level in China is way too high. However, its super-sized bubble is bursting, and millions of Chinese are struggling to afford a place to live, so Xi doesn't know how he could snatch another term. He doesn't even know how to flirt with the idea that he could run for it.

China is ruled by a one-party authoritarian dictatorship. The state is entirely under the monopolistic control of a single ruling political party called the Chinese Communist Party. The government continued to implement far-reaching policies that severely restricted its people's freedoms. It is not a government of the people, by the people, for the people. It was. Before Xi's administration, the government paid attention to public opinion and, to a certain extent, sought to lay the foundation for protecting and providing for the public, drawing on lessons from history because the dynasties didn't protect citizens from violence, and the worst vicissitudes of life eventually fell. History repeats itself in historical recurrence.

When we take a close look at the historical dynamics of the Chinese dynasties, we can see that the foundation of a just government can be built well, rule goes well, and the leader also behaves well until the emperor dies, the generations then change, and one ruler gets replaced after another, and corruption gradually rises to the surface. The power eventually fails to garner public sympathy. In the end, the population voices dissatisfaction and anger as they turn to revolt. The dynasty then dies. These recurring upheavals serve as a stark reminder.

"Romance of the Three Kingdoms," the most famous historical fiction book of all time, begins with the fall of the Han Dynasty and a series of protests and uprisings known as the "Yellow Turban Rebellion" across China. Learning from this history, the previous leaders before Xi Jinping (also from the Chinese Communist Party, though) understood and managed public sentiment, even though it was never truly democratic. Cultural history always provides valuable lessons for today's society. If stagflation warning signs keep emerging in the Chinese economy, it is scary to wonder when people will blow up.

Meanwhile, in 2027, Communist Party members will vote at the Communist Party Congress to decide whether Xi Jinping will be nominated to run for a fourth term as their president. The dictator has complete control over all authorities in China, but that does not mean he can hold the pent-up anger. Political outrage can be expressed through voting, and Xi definitely needs to do something to get re-elected for the next term.

So, what if China invaded Taiwan to cover up its recession…?

Chapter 4

The terrifying ending

"The new page will begin in 2027, a few months before the Communist Party Congress in China. All of a sudden, Chinese naval warships invaded the waters near the Senkaku Islands, just kilometers away from Taiwan. The island will then be surrounded. The moment the news hit, people all over the world were shocked, and the roar of the fleet's guns rang out in unison, signaling the start of war... And it is the beginning of China's military plans to defeat and invade Taiwan.

The terrifying scenes that unfolded in Taiwan were witnessed by people all over the world, and the United Nations immediately called for an emergency session. Despite some opposition from Russia, North Korea, and other countries, the UN passed a resolution by a majority in favor, condemning China and halting its army contracts.

Shortly after this, the US, the EU, and other Western countries, except for those with close economic ties to China, decided to unite and impose strong economic sanctions on China, one after another, freezing the assets of Beijing and its key figures worldwide and halting trade. However, despite the active intervention of the US and other Western nations, the fierce battle between the two superpowers around the Taiwan Strait will degenerate into a mere spectacle of meaningless, long wars …

This means, of course, a halt to Western exports of cars and other goods to China, as well as a halt to imports of car batteries and other EV components from China.

Western automakers will then have no choice but to abandon EV production..."

Afterthoughts

As for the issue of EVs in the Sino-US trade war, this is only the beginning. China's need for rare earths is not limited to the automotive industry; they are now indispensable across a range of fields, including semiconductors and military weapons. We must immediately stop using these rare earths, and governments and companies around the world must prioritize developing alternative technologies to replace them. We don't have the evidence to back up the peaceful story that China will never invade Taiwan. China, after all, is not that simple.

There is no playback or pause button for this story, no matter how influential any corporation is. From a macro perspective, pro-China political parties are losing elections one after another in the United States and other countries worldwide, while anti-China pledges are gaining popularity. Aside from the fragile economic thaw between the US and China, Beijing might eventually halt all rare-earth exports in the near future.

The deadliest mistake in Biden's administration was the misconception that China would be over the moon if we, Western coun-

peaceful coexistence and mutual prosperity.

Russia is different from China, however. Russians have moral goodness, meaning they don't take away others' property for their own sake. The Russian government cares for its people. It won't put its people in hot water. Still, if you make them feel sidelined for your own good, you pay for what you choose.

No matter how dexterous China is, mistakes were made under President Trump's aggressive strategy to curb Beijing's economic dominance. Through all these tariffs and supply chain disruptions of rare earth elements, we came to understand that the true, real problem lies in the stable supply of rare earths. Our world needs those elements, and surely it will wisely seek alternatives in materials and technologies.

Even without a Taiwan invasion, electric vehicles will still die in 2027 in a different sense.

tries, purchased a large quantity of rare earths from them. Xi, the terrible and wicked leader, has no clue how to create sustainable solutions and drive economic growth. What he only knows is how to put Western countries into trouble. For this reason, we must be well-prepared when Beijing stops exporting rare earths, leaving only the end products.

The problem is not something for the West only. The BRICS (a coalition formed by China, Russia, India, Brazil, and South Africa to counter advanced Western countries) must also get prepared to face the same fate with courage. No matter how furious you might be with the United States and the West, you must never let that animosity outweigh the alertness to China and Russia. This is just not worth it. The people in Japan have already experienced that firsthand.

Japan was defeated in World War II in 1945. Russia occupied the Northern Territories in the north of the country, and the United States occupied Okinawa in the southwest. And now, 80 years after the war, what are these territories like? Okinawa was returned to Japan safely, but the Northern Territories remain under Russian control. The United States and other Western countries would never take anything from anyone. They understand the concept of

Author profile

Masafumi OYAMA

1972 Born in Aichi Prefecture, Japan; later graduated from Wako University

2023 Became an active economist

Previous publications include “Welcome 30 million immigrants if you want to rebuild your finances”, “Short-term rate hikes, long-term rate cuts”, and “Dismantling the Ministry of Finance”.